STEM CAREERS

GEOLOGIST

by Nikole Brooks Bethea

pogo

Ideas for Parents and Teachers

Pogo Books let children practice reading informational text while introducing them to nonfiction features such as headings, labels, sidebars, maps, and diagrams, as well as a table of contents, glossary, and index.

Carefully leveled text with a strong photo match offers early fluent readers the support they need to succeed.

Before Reading

- "Walk" through the book and point out the various nonfiction features. Ask the student what purpose each feature serves.
- Look at the glossary together. Read and discuss the words.

Read the Book

- Have the child read the book independently.
- Invite him or her to list questions that arise from reading.

After Reading

- Discuss the child's questions. Talk about how he or she might find answers to those questions.
- Prompt the child to think more. Ask: Do you know anyone who works as a geologist? What projects has he or she been involved in? Do you have any interest in this kind of work?

Pogo Books are published by Jump!
5357 Penn Avenue South
Minneapolis, MN 55419
www.jumplibrary.com

Library of Congress Cataloging-in-Publication Data

Names: Bethea, Nikole Brooks, author.
Title: Geologist / by Nikole Brooks Bethea.
Description: Minneapolis, MN: Jump!, Inc., [2019]
Series: STEM careers | Audience: Ages 7-10.
Includes bibliographical references and index.
Identifiers: LCCN 2018014522 (print)
LCCN 2018018842 (ebook)
ISBN 9781641281874 (ebook)
ISBN 9781641281867 (hardcover: alk. paper)
Subjects: LCSH: Geology–Vocational guidance–Juvenile literature. | Geologists–Juvenile literature.
Classification: LCC QE29 (ebook) | LCC QE29 .B478 2018 (print) | DDC 550.23–dc23
LC record available at https://lccn.loc.gov/2018014522

Editors: Jenna Trnka and Susanne Bushman
Designer: Michelle Sonnek
Content Consultant: Sandra Feher, M.S.G.E.

Photo Credits: wawritto/Shutterstock, cover; Phil Hill Geo Pics/Alamy, 1; HStocks/Shutterstock, 3 (child); Bragin Alexey/Shutterstock, 3 (rock); Nagel Photography/Shutterstock, 4; Checubus/Shutterstock, 5; Keith Douglas/Alamy, 6-7; Roger Ressmeyer/Getty, 8-9; Zeljko Radojko/Shutterstock, 10 (left); wavebreakmedia/Shutterstock, 10 (right); Kanok Sulaiman/Shutterstock, 11; l3l3oysky/Shutterstock, 12-13 (background); Chris Sattlberger/Science Source, 12-13 (foreground); Biosphoto/Superstock, 14-15; Keith Douglas/Superstock, 16-17; Thaweewong Vichaiururoj/Shutterstock, 18; RichVintage/iStock, 19 (foreground); Picsfive/Shutterstock, 19 (background); fotoVoyager/iStock, 20-21; anaken2012/Shutterstock, 23.

Printed in the United States of America at Corporate Graphics in North Mankato, Minnesota.

TABLE OF CONTENTS

CHAPTER 1

STUDYING THE PAST

Look at Mount Rushmore. Do you see presidents? A geologist sees **granite.**

What do you notice about the Grand Canyon? Its depth? Geologists see **erosion**. They see Earth's history.

Geologists study Earth's materials. And how Earth has changed over time. How? One way is studying rock. Rock deep in the Earth tells them about the past.

What else? Rock can give details about past **climates**. This tells us how today's climate is changing.

DID YOU KNOW?

Geologists study **fossils** found in rock layers. They may find an ocean animal fossil. This can mean the area was once an ocean.

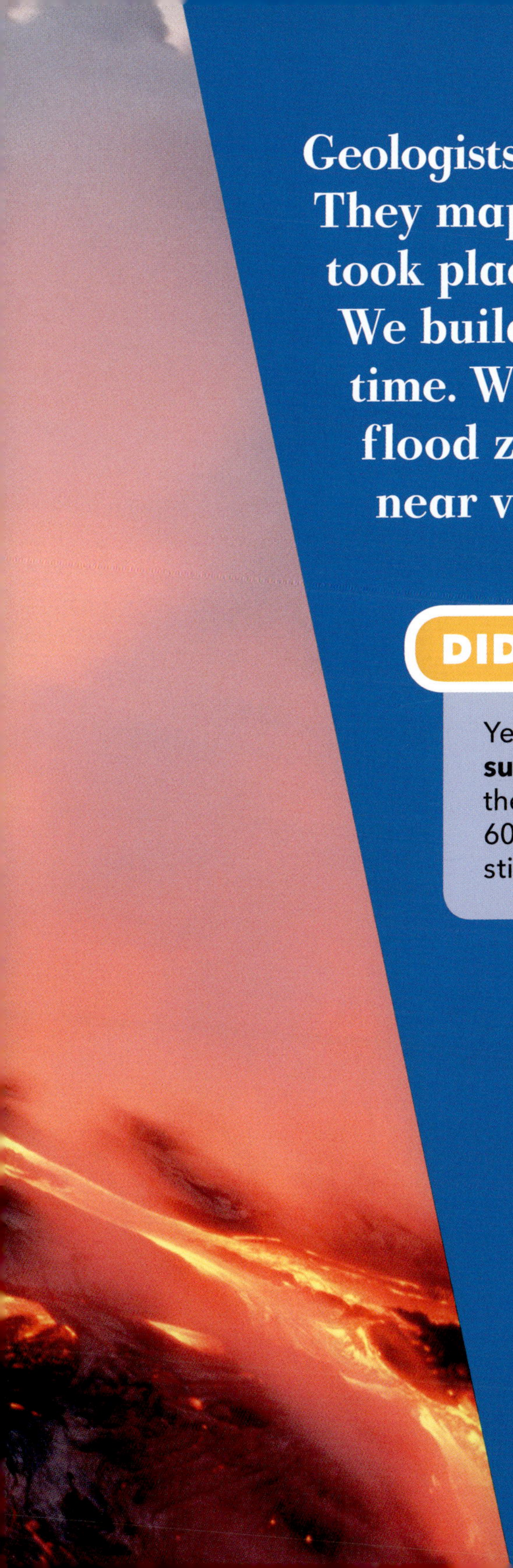

Geologists study **natural disasters.** They map the areas where they took place. We learn from them. We build in safer places next time. We avoid earthquake and flood zones. We do not build near volcanoes.

DID YOU KNOW?

Yellowstone National Park has a **supervolcano** under it. When was the last major eruption? More than 600,000 years ago. But geologists still keep a close eye on it.

CHAPTER 2

WHAT DO THEY DO?

Some geologists search for oil and gas. Where? Under both land and sea. They drill. They collect samples. They test them for oil and gas.

They find the best way to bring oil and gas to the surface. Then we use it for **energy**. It fuels our cars. It is made into electricity. It powers our homes.

Geologists use **computer models**. These show what is underground. They use them to find minerals, rocks, and gems. Then they plan safe ways to mine them.

DID YOU KNOW?

Dynamite can be used to extract **resources**. Geologists play a role. They make sure the blast will not damage anything nearby. They make sure mining will not **pollute** groundwater.

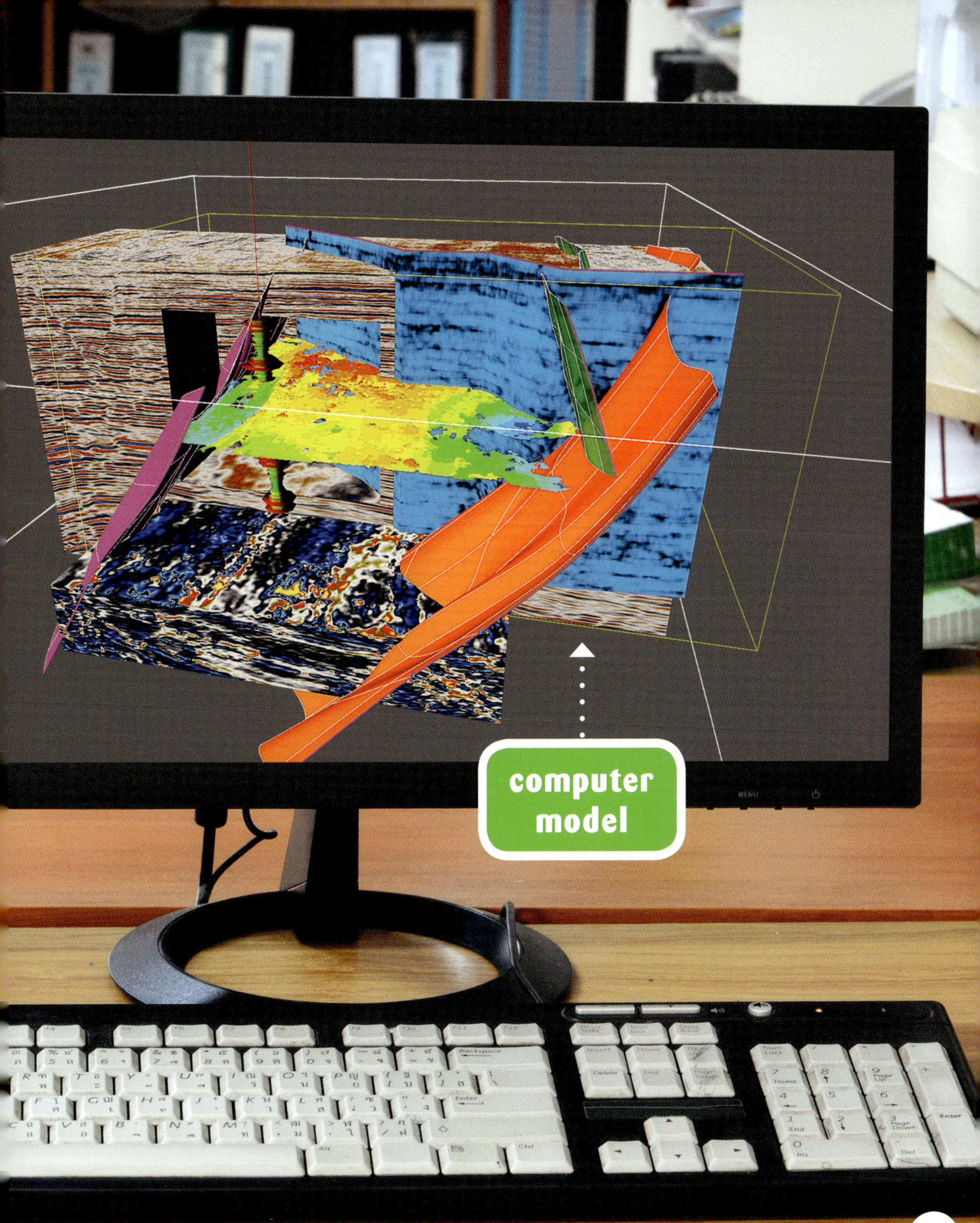
computer model

landfill

They work to keep Earth clean. They look for ways to safely remove waste. This must be done without polluting land or water. Or they find ways to contain it. An example is a **landfill**.

Geologists work on environmental cleanups. They create plans to test soil and groundwater.

They spend a lot of time working outside. This is called fieldwork. They take **core samples**. Why? To know what is underground. They **bore** a hole. How? They drill into the ground with a hollow tube. Soil and rock collect inside the tube. Geologists then identify the rock and soil types in the sample. They record the findings.

TAKE A LOOK!

Findings are recorded. In what? A **boring log**. What does this one tell us about the soil sample?

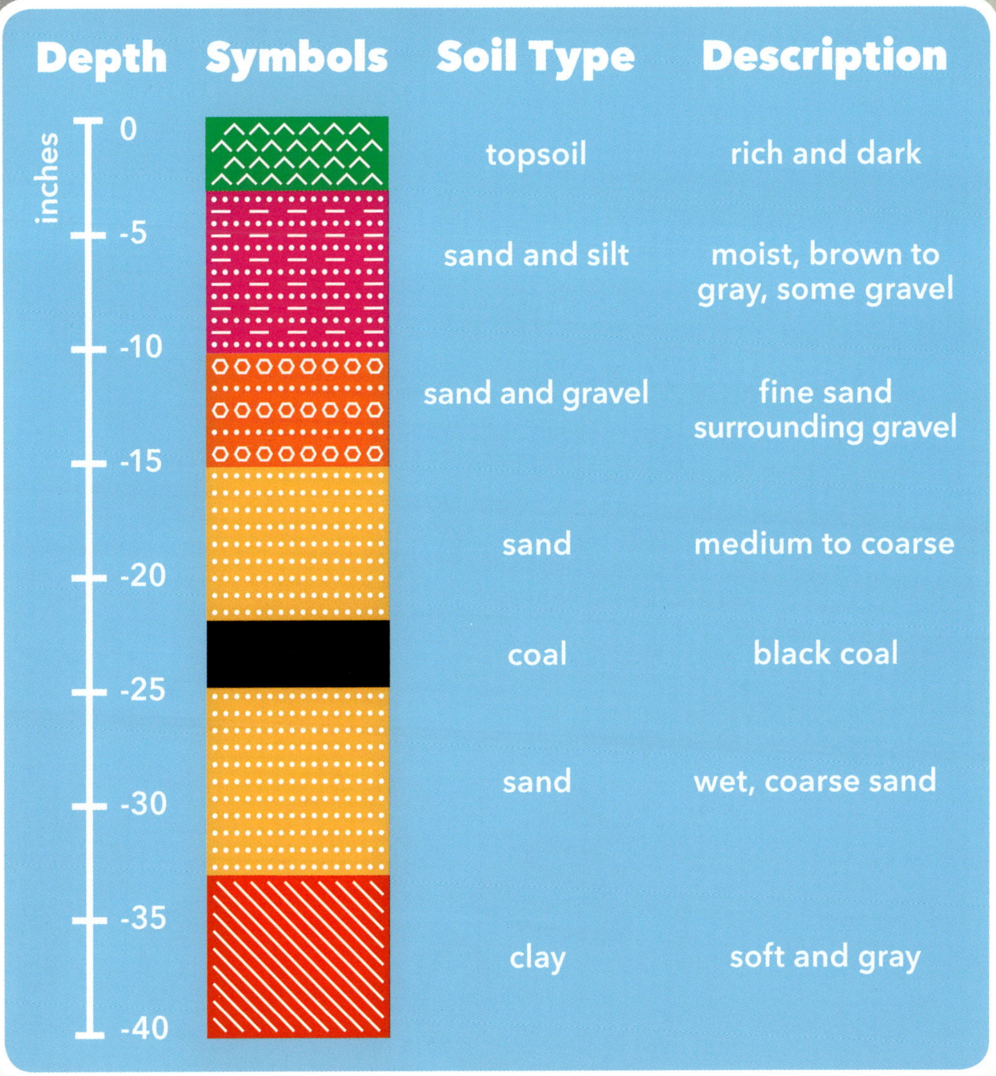

CHAPTER 3

BECOMING A GEOLOGIST

Do you want to be a geologist? You must love the outdoors. And have observation skills. Start studying geology. What else? **Biology**. **Chemistry**. Physical science.

Math and writing skills are important, too. Why? For calculating. And writing reports. You'll need a college degree. And you must pass a test to get licensed.

Do you like rocks? And exploring the outdoors? You could be a geologist! You may find a new oilfield. You could make an important discovery.

DID YOU KNOW?

To work as a geologist, you need STEM skills. What does STEM stand for? Science. Technology. Engineering. Math. STEM careers are in demand. They pay well, too.

ACTIVITIES & TOOLS

TRY THIS!

CORE SAMPLE ACTIVITY

Collect a core sample! Record your results in a boring log.

What You Need:

- slices of different breads (white, wheat, etc.)
- different spreads (plain and crunchy peanut butter, grape jelly, etc.)
- small, dry filling pieces (raisins, chocolate chips, cereal, etc.)
- large tube noodle, such as manicotti
- plastic knife
- paper towel
- sheet of paper
- drinking straw
- colored pencils

1. **Place one slice of bread on the paper towel. This represents a layer of soil.**
2. **Spread a filling, such as peanut butter or jelly, on the bread. This represents another soil layer, such as clay.**
3. **Sprinkle one of the dry fillings over the spread layer. This represents stones or shells that are deposited by running water or melting snow.**
4. **Continue layering bread, spread, and dry pieces, finishing with bread on top.**
5. **Press the tube of manicotti down into your geology sandwich. Lift out the tube.**
6. **Gently press the straw into the manicotti tube to press the sandwich core out onto the paper towel.**
7. **Draw a picture of your core sample. Use different colors and symbols to identify each layer. This is your soil boring log.**

GLOSSARY

biology: The study of life and living things.

bore: To make a hole in something with a drill or similar tool.

boring log: A written record of information about a soil sample bored or drilled from the Earth.

chemistry: A science that deals with the structure of substances and the changes they go through.

climates: Weather conditions typical of an area over a long period of time.

computer models: Representations of real-world systems or processes on a computer, used to help calculate and predict.

core samples: Cylinder-shaped sections taken from a naturally-occurring substance, such as rock or soil.

energy: Usable power.

erosion: The process of wearing away from water, wind, heat, or ice.

fossils: The remains, traces, or prints of plants or animals from the past that are preserved in rock.

granite: A hard igneous rock used for building structures and monuments.

landfill: A site where waste is buried between layers of Earth.

natural disasters: Natural events, such as floods, earthquakes, and hurricanes, that cause great damage or loss of life.

pollute: To make the air, water, or soil unsafe to use.

resources: Supplies of something.

supervolcano: An unusually large volcano with the potential to produce an eruption that would cause major effects on Earth's climate and ecosystem.

INDEX

TO LEARN MORE

Finding more information is as easy as 1, 2, 3.

1. Go to www.factsurfer.com
2. Enter "geologist" into the search box.
3. Click the "Surf" button to see a list of websites.